INTRODUCTION

This booklet by the Astronomy Corre  e,
provides a convenient guide to seeing t e.
The twelve monthly charts on pp. 4–2(in
the evening. The charts have been draw ut
may be used for any part of the Briti he
visibility of the planets and phases of t se
are tabulated on page 29 along with notes on the principal ..... ier
notes on the more interesting events in 2001 appear on page 31. A detailed explanation
of the astronomical terms used and how the various phenomena arise can be found in
the fully illustrated *The Times Night Sky Companion*.

The Changing Aspect of the Night Sky

From our position on the surface of the Earth, the stars appear to lie on the inside of a
spherical surface, called the celestial sphere. Because the stars are so far away their
directions remain essentially unchanged when seen from different parts of the Earth's
orbit. The diagram on the inside cover *(see* left) shows one such direction, indicated
by the arrow pointing to the First Point of Aries. The stars that lie behind the Sun as
seen from the Earth at the beginning of April each year will, by October, be in the
opposite part of the sky to the Sun and be due south at midnight. The arrow points to
Pisces (where the First point of Aries now lies) and looking at the October chart Pisces
is indeed in the southern sky but it does not appear at all on the April chart, being
behind the Sun and in the daytime sky. More generally, a line from the Sun through the
position of the Earth points towards the stars seen near the lower centre of the chart for
that month. Remembering this, one can relate the positions of the other planets in their
orbits to where they will be in the sky, though they will not always be on the monthly
chart, being too near to the Sun in direction.

Time of Observation and Location

Greenwich Mean Time (GMT but also known as Universal Time) is used throughout
this booklet. When in force, British Summer Time (BST) is 1 hour ahead of GMT, e.g.
23h BST is 22h GMT. Strictly speaking, the charts are only correct in the stars they
show above the horizon for an observer near London (Greenwich). As one moves north
fewer stars appear above the southern horizon. Movement east or west along the same
latitude does not alter what stars can be seen, but only when they can be seen.

Using the Charts

The charts show the brighter stars above the horizon for London at 23h (11pm) at the
beginning, 22h (10pm) in the middle and 21h (9pm) at the end of each month. The
stars rise four minutes earlier each night or two hours earlier each month, being back
in their same positions at the same time after a year. Thus, for instance, the aspect of
the heavens at 23h on 1 April is the same as on 1 May at 21h or 1 March at 01h. By

remembering this rule, the chart applicable to any hour throughout the year may be found. This rule does not apply to the Moon and planets. The charts show the whole sky visible at one time with the zenith, the point directly overhead, at the centre of the chart. Note that the Pole Star (Polaris) occupies the same position on every chart being close to one of the two points around which the whole star sphere appears to revolve. It is easily found in relation to Ursa Major at all times of the year and is useful in defining due north. Ursa Major's seven brightest stars form the Plough. The end two stars (the Pointers) are always in line with Polaris. If the observer faces south with the Pole Star to his back and the appropriate chart held up as one would read the booklet, the constellations depicted above the southern horizon should be to the front, with the eastern aspect to the left and western horizon to the right.

Explanatory Notes on Terms Used

The Moon – the phase and position are given for about 22h on every other day when it is above the horizon at that time. The average time between like phases (e.g. full to full) is 29.5 days, 2 days longer than it takes to return amongst the same stars. It moves eastwards by its own diameter every hour.

The Planets – are shown in the position they occupy about the middle of the month unless otherwise indicated, and for Venus and Mars an arrow shows by its length the movement during the month. Planets crossing the meridian (i.e. due south) before midnight are said to be evening stars while those crossing the meridian after midnight are morning stars. A planet is in opposition to the Sun when it is in the opposite part of the sky to the Sun and therefore due south at midnight. (Mercury and Venus can never be at opposition.) It is then at its closest and brightest for that year. For a few weeks on either side of opposition, motion among the stars, instead of being from west to east as usual, is from east to west and is called retrograde. At the turning points, where motion is reversed, the planet is said to be stationary. A planet coming in line with the Earth and the Sun is said to be in superior conjunction with the Sun if it lies beyond the Sun but at inferior conjunction if it lies between the Earth and the Sun. Only Mercury and Venus can be at inferior conjunction. Planets can also be in conjunction with others when close in the sky. Mercury and Venus are said to be at greatest elongation when at their greatest apparent distance from the Sun, either east (evening) or west (morning). They can never be high in the sky late at night. Mercury is not observable in a dark sky from the British Isles and may require binoculars. It is always too near the sun to be included on the monthly charts. Uranus is visible at times to the naked eye but will probably require binoculars for identification. Neptune always requires optical aid. Pluto requires a moderate-sized telescope and is not mentioned in the monthly notes. Opposition in 2001 is on 4 June, the 14th magnitude planet being in Ophiuchus.

ECLIPSES IN 2001

9 JANUARY

This total eclipse of the Moon will be visible from W Australia, Asia, Africa, Europe (including the British Isles), Greenland, N Canada and N Alaska. The Moon enters the umbra at 18h 42m, totality begins 19h 50m, totality ends 20h 52m and the Moon leaves the umbra at 22h 00m.

21 JUNE

The track of this total eclipse of the Sun passes from the S Atlantic Ocean, over parts of Angola, Zambia, Zimbabwe, Mozambique, Madagascar and into the Indian Ocean. In SE Africa totality occurs from 13h UT (mid-afternoon local time) and lasts about 3.5 minutes. A partial eclipse will be seen over much of E South America, the S Atlantic Ocean, central and southern Africa and the SW Indian Ocean.

5 JULY

This partial eclipse of the Moon will be visible over Antarctica, New Zealand, Australia, eastern and central Asia, E Africa, Hawaiian Islands and the central Pacific Ocean.

14 DECEMBER

This annular eclipse of the Sun will be visible as a partial eclipse over the Pacific Ocean, NW South America, Central America, USA, and W and S Canada. The much narrower path from which the annular eclipse can be seen reaches landfall in Central America, just before sunset.

30 DECEMBER

This penumbral eclipse of the Moon can be seen from Australasia, much of Asia, extreme N Europe, Greenland, the Americas and the Pacific Ocean. In penumbral eclipses the Moon passes only through the outer, lighter shadow of the Earth and the Moon is only slightly darkened. Such eclipses often go unnoticed.

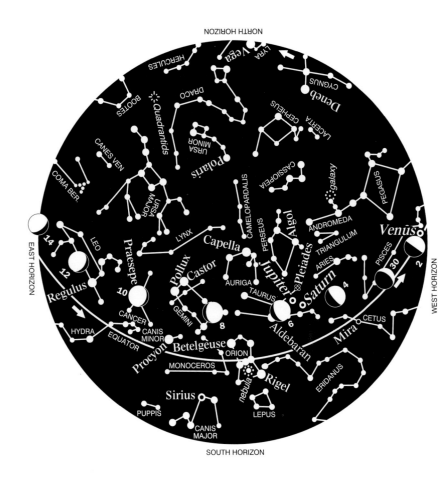

JANUARY 1, 23h (11pm)

The aspect of the sky (apart from the Moon and Planets) will be approximately the same in other months at the following times:

October 1, 05h: November 1, 03h: December 1, 01h: February 1, 21h: March 1, 19h.

The time in these notes is that of the Greenwich meridian.

4

JANUARY

The Planets

MERCURY begins the year in the evening sky fading from -1 to 0.0 magnitude during January. Setting about 1.5 hours after the Sun, it will be visible in the SW from the 20th.

VENUS is a brilliant -4.5 magnitude evening star, setting after 21h by the 31st. Greatest eastern elongation (47 degrees) on the 17th. Moon nearby on the 28th.

MARS moves from Virgo into Libra in early January, the 1.2 magnitude red planet rising about 02h by the 31st. Moon nearby on the 18th.

JUPITER is in Taurus, stationary on the 25th. A bright -2.6 magnitude, it sets about 03h by end month. Moon nearby on the 6th.

SATURN is in Taurus, stationary on the 25th. The 0.0 magnitude ringed planet sets about 02h by the 31st. Moon nearby on the 5th.

URANUS is in Capricornus setting soon after the Sun by the 31st.

NEPTUNE is also in Capricornus, in conjunction with the Sun on the 26th.

The Moon

First quarter 2d 23h
Full Moon 9d 20h
Last quarter 16d 13h
New Moon 24d 13h

Eclipse on the 9th: *see* page 3
The Earth: at perihelion 4d 09h (147 million km)

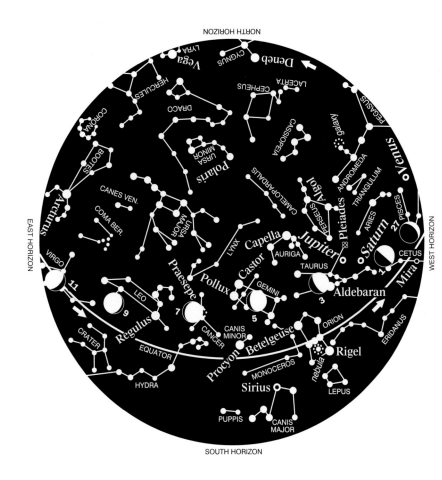

FEBRUARY 1, 23h (11pm)

The aspect of the sky (apart from the Moon and Planets)
will be approximately the same in other months at the
following times:

**November 1, 05h: December 1, 03h: January 1, 01h:
March 1, 21h: April 1, 19h.**

The time in these notes is that of the Greenwich meridian.

FEBRUARY

The Planets

MERCURY may be visible low in the SW during the first days of the month. At inferior conjunction on the 13th, it becomes a morning object but is too near the Sun for observation.

VENUS sets after 21h during February, reaching greatest brilliancy (-4.6 magnitude) on the 22nd. The position shown on the chart is for the 28th.

MARS moves from Libra into Scorpius late in the month. Rising before 02h Mars, now 0.8 magnitude, is brightening slowly as it approaches opposition in June. Moon nearby on the 15th.

JUPITER is in Taurus and below the Pleiades star cluster. Moon to the south on the 2nd.

SATURN is in Taurus throughout the year. Moon nearby on the 1st and 2nd.

URANUS is in conjunction with the Sun on the 9th, then becoming a morning object.

NEPTUNE rises about an hour before the Sun by the 28th.

The Moon

First quarter 1d 14h
Full Moon 8d 07h
Last quarter 15d 03h
New Moon 23d 08h

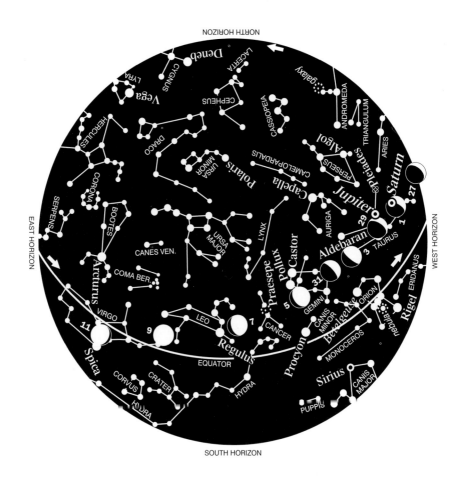

MARCH 1, 23h (11pm)

The aspect of the sky (apart from the Moon and Planets) will be approximately the same in other months at the following times:

December 1, 05h: January 1, 03h: February 1, 01h: April 1, 21h.

The time in these notes is that of the Greenwich meridian.

MARCH

The Planets

MERCURY is at greatest western elongation (27 degrees) on the 11th but too far south to be observable from the British Isles. VENUS sets about 21h on the 1st, 3.5 hours after sunset. It closes rapidly with the Sun to be at inferior conjunction on the 30th, after which Venus will be a morning star for the rest of the year.

MARS rises before 01h by the 31st when the planet will be -0.2 magnitude. Passing 5 degrees north of Antares in Scorpius (1.0 magnitude) on the 4th, Mars moves into Ophiuchus in early March. Moon nearby on the 16th.

JUPITER is -2.1 magnitude by the 31st when it sets before 0h. Moon nearby on the 2nd and 2 degrees south on the 29th.

SATURN is 0.2 magnitude and sets before 23h by the 31st. Moon nearby on the 1st and 28th.

URANUS rises about 04h 30m by the 31st. Although bright enough at 5.7 magnitude to be seen with the unaided eye in a clear dark sky, binoculars or a small telescope are usually needed to locate it.

NEPTUNE rises by 03h 30m by the 31st. Neptune is 8 magnitude and so more difficult to identify than Uranus. To show their tiny discs, a telescope with a magnification of 80 times or more is required, but they may be identified also by plotting their slow motion from night to night among nearby fixed stars.

The Moon

First quarter 3d 02h
Full Moon 9d 17h
Last quarter 16d 21h
New Moon 25d 01h

The Earth: Spring Equinox 20d 14h

APRIL 1, 23h (11pm)

The aspect of the sky (apart from the Moon and Planets) will be approximately the same in other months at the following times:

December 1, 07h: January 1, 05h: February 1, 03h: March 1, 01h: May 1, 21h.

The time in these notes is that of the Greenwich meridian.

APRIL

The Planets

MERCURY is at superior conjunction on the 23rd, becoming an evening star to be visible in May.

VENUS is in morning twilight, rising about an hour before the Sun.

MARS rises before 0h by the 30th, passing into Sagittarius in mid-month. Moon close by on the 13th.

JUPITER passes 5 degrees north of Aldebaran in Taurus on the 16th, setting about 22h by the 30th. Moon nearby on the 25th–26th.

SATURN is in Taurus, setting by 21h by the end of April. Moon nearby on the 25th.

URANUS is in Capricornus and rises after 02h by the 30th. Moon nearby on the 17th.

NEPTUNE is also in Capricornus rising before 02h by the 30th. Moon nearby on the 16th.

The Moon

First quarter 1d 11h
Full Moon 8d 03h
Last quarter 15d 16h
New Moon 23d 15h
First quarter 30d 17h

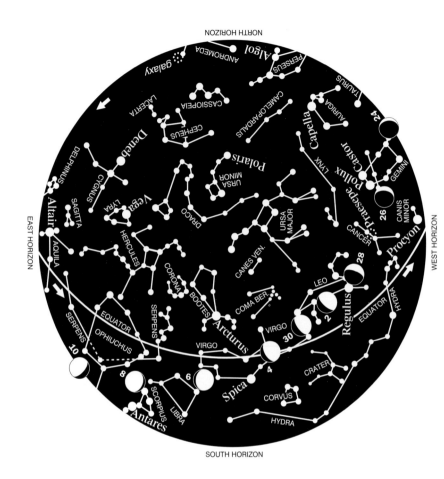

MAY 1, 23h (11pm)

The aspect of the sky (apart from the Moon and Planets)
will be approximately the same in other months at the
following times:

**January 1, 07h: February 1, 05h: March 1, 03h:
April 1, 01h: June 1, 21h.**

The time in these notes is that of the Greenwich meridian.

MAY

The Planets

MERCURY will be visible in the western evening sky from early May to its greatest eastern elongation (22 degrees) on the 22nd. The best apparition of the year, Mercury fades from -1.5 on the 1st to 0.0 magnitude soon after mid-month when it sets 2 hours after the Sun. Saturn to the south on the 7th, Jupiter to the south on the 16th and Moon to the south on the 24th.

VENUS is in bright morning twilight throughout May but might be glimpsed near the northeast horizon. It reaches greatest brilliancy -4.5 magnitude on the 4th. Moon nearby on the 19th.

MARS rises by 22h on the 31st when it reaches -2.0 magnitude. It is stationary on the 11th and then retrogrades, moving westwards against the stars of Sagittarius. Moon nearby on the 10th–11th.

JUPITER is -1.9 magnitude and in Taurus, setting by 21h by end May. Moon nearby on the 24th.

SATURN is in conjunction with the Sun on the 25th and will not be observable.

URANUS is stationary on the 29th, rising soon after 0h by the 31st. Moon to the south on the 15th.

NEPTUNE is stationary on the 11th, rising before 0h by the end of May. Moon to the south on the 14th.

The Moon

Full Moon 7d 14h
Last quarter 15d 10h
New Moon 23d 03h
First quarter 29d 22h

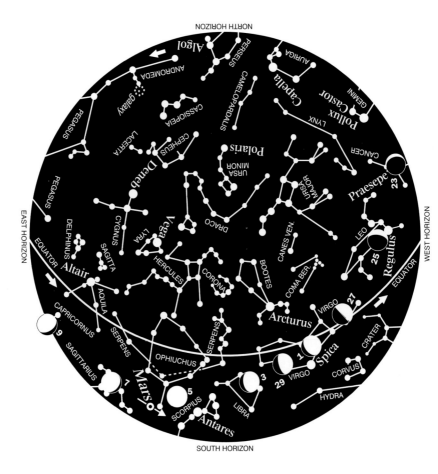

JUNE 1, 23h (11pm)

The aspect of the sky (apart from the Moon and Planets) will be approximately the same in other months at the following times:

February 1, 07h: March 1, 05h: April 1, 03h:
May 1, 01h: July 1, 21h.

The time in these notes is that of the Greenwich meridian.

JUNE

The Planets

MERCURY reaches inferior conjunction on the 16th and then becomes a morning object. It will not be observable in June.

VENUS is at greatest western elongation (46 degrees) on the 8th, and rises 2 hours before the Sun by the 30th. A brilliant -4.3 magnitude it will be visible in the northeastern twilight. Moon nearby on the 18th.

MARS moves westwards from Sagittarius back into Ophiuchus in early June, coming to opposition on the 13th. At its closest to the Earth on the 21st it reaches -2.4 magnitude, approaching Jupiter's maximum brightness. The planet's low altitude will severely hamper telescopic observers in mid and high northern latitudes. Moon nearby on the 5th and 6th.

JUPITER is in conjunction with the Sun on the 14th and will not be observable this month.

SATURN is in Taurus, rising nearly 2 hours before the Sun by the 30th, but still in twilight. Moon nearby on the 20th.

URANUS is in Capricornus rising after 22h by the 30th. Moon nearby on the 11th.

NEPTUNE is also in Capricornus rising before 22h by the 30th. Moon nearby on the 10th.

The Moon

Full Moon 6d 02h
Last quarter 14d 03h
New Moon 21d 12h
First quarter 28d 03h

Eclipse on the 21st: *see* page 3
The Earth: Summer Solstice on 21d 08h

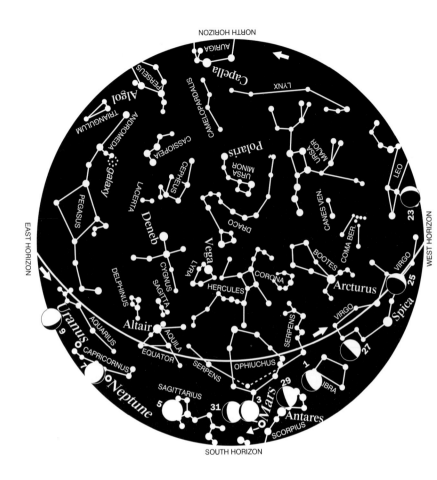

JULY 1, 23h (11pm)

The aspect of the sky (apart from the Moon and Planets) will be approximately the same in other months at the following times:

**April 1, 05h: May 1, 03h: June 1, 01h:
August 1, 21h: September 1, 19h.**

The time in these notes is that of the Greenwich meridian.

JULY

The Planets

MERCURY is a morning object reaching greatest western elongation (21 degrees) on the 9th. Rising 1.5 hours before the Sun on the 15th, though brighter than 0.0 magnitude it will be in bright twilight and difficult to observe. Close to Jupiter on the 12th–13th.

VENUS rises in the northeast soon after 01h by the 31st and will be a conspicuous morning star until November. North of Aldebaran and close to Saturn on the 15th and Moon nearby on the 17th.

MARS is stationary on the 19th and then resumes direct or eastwards motion against the stars and will have faded to -1.5 magnitude by the 31st. Moon to the north on the 3rd and 30th.

JUPITER at -2.0 magnitude moves from Taurus into Gemini in mid-July. Close to the Moon on the 19th.

SATURN is in Taurus passing north of Aldebaran about the13th and rising about 0h by end month. Moon nearby on the 17th.

URANUS rises at sunset by the 31st. Moon nearby on the 8th.

NEPTUNE is at opposition on the 30th. Moon nearby on the 7th.

The Moon

Full Moon 5d 15h
Last quarter 13d 19h
New Moon 20d 20h
First quarter 27d 10h

Eclipse on the 5th: *see* page 3
The Earth: at aphelion 4d 14h (152 million km)

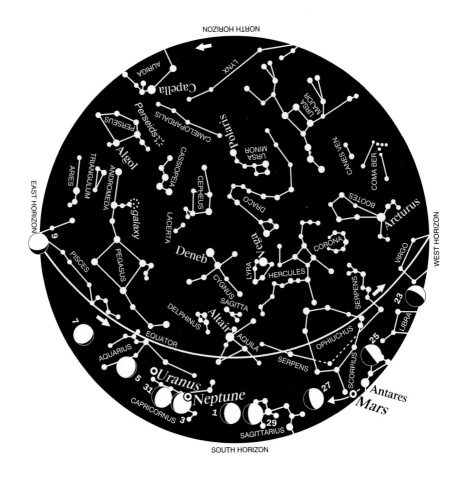

AUGUST 1, 23h (11pm)

The aspect of the sky (apart from the Moon and Planets) will be approximately the same in other months at the following times:

June 1, 03h: July 1, 01h: September 1, 21h: October 1, 19h: November 1, 17h.

The time in these notes is that of the Greenwich meridian.

AUGUST

The Planets

MERCURY is at superior conjunction on the 5th and then becomes an evening object but will be in too bright a sky for observation for the rest of the year.

VENUS rises 3 hours before the Sun, a brilliant morning star. Jupiter to the north on the 6th and Moon nearby on the 16th–17th.

MARS is in Ophiuchus setting soon after 22h by end August when it will have faded to -1.0 magnitude. Moon to the north on the 27th

JUPITER is in Gemini, rising at midnight by the 31st. Moon to the north on the 16th.

SATURN is in Taurus, rising about 22h 30m by the 31st. Moon to the south on the 14th.

URANUS is in Capricornus and at opposition on the 15th. Moon nearby on the 4th and 31st.

NEPTUNE is in Capricornus and sets about 02h 30m by the 31st. Moon nearby on the 3rd and 30th.

The Moon

Full Moon 4d 06h
Last quarter 12d 08h
New Moon 19d 03h
First quarter 25d 20h

SEPTEMBER 1, 23h (11pm)

The aspect of the sky (apart from the Moon and Planets) will be approximately the same in other months at the following times:

July 1, 03h: August 1, 01h: October 1, 21h: November 1, 19h: December 1, 17h.

The time in these notes is that of the Greenwich meridian.

SEPTEMBER

The Planets

MERCURY is an evening object at greatest eastern elongation (27 degrees) on the 18th but will not be observable.

VENUS is a brilliant morning star rising 3 hours before the Sun. Venus north of Regulus on the 20th. Moon to the north on the 15th.

MARS passes from Ophiuchus into Sagittarius in early September, setting before 22h by the 30th when the brightness will have fallen to -0.4 magnitude. Moon to the north on the 24th.

JUPITER is in Gemini, rising soon after 22h in late September. Moon nearby on the 12th.

SATURN is 0 magnitude, in Taurus and stationary on the 27th. It rises by 20h 30m on the 30th. Moon nearby on the 10th.

URANUS is in Capricornus, setting by 02h by the 30th. Moon nearby on the 1st and 28th.

NEPTUNE is in Capricornus setting soon after midnight by the 30th. Moon nearby on the 27th.

The Moon

Full Moon 2d 22h
Last quarter 10d 19h
New Moon 17d 10h
First quarter 24d 10h

The Earth: Autumn Equinox on 22d 23h

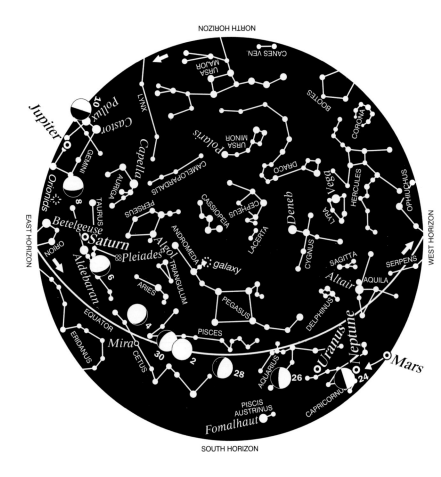

OCTOBER 1, 23h (11pm)

The aspect of the sky (apart from the Moon and Planets) will be approximately the same in other months at the following times:

August 1, 03h: September 1, 01h: November 1, 21h: December 1, 19h: January 1, 17h.

The time in these notes is that of the Greenwich meridian.

OCTOBER

The Planets

MERCURY is at inferior conjunction on the 14th and then becomes a morning star reaching greatest western elongation (19 degrees) on the 29th but it will remain too near the horizon for observation.

VENUS closes with the Sun during October, rising only 2 hours before sunrise by the 31st. Moon to the north on the 15th.

MARS is 0 magnitude and moves into Capricornus in late October. Moon close by on the 23rd.

JUPITER is in Gemini and -2.4 magnitude. It rises by 20h 30m on the 31st. Moon close and to the north on the 9th.

SATURN rises soon after 18h by end October. Moon close by on the 7th.

URANUS is stationary on the 31st and when it sets before midnight. Moon nearby on the 25th.

NEPTUNE is stationary on the 17th and sets about 22h 30m by the 31st. Moon nearby on the 24th.

The Moon

Full Moon 2d 14h
Last quarter 10d 04h
New Moon 16d 19h
First quarter 24d 03h

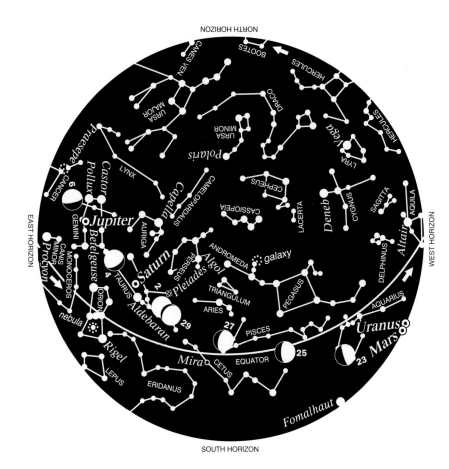

NOVEMBER 1, 23h (11pm)

The aspect of the sky (apart from the Moon and Planets) will be approximately the same in other months at the following times:

September 1, 03h: October 1, 01h: December 1, 21h: January 1, 19h: February 1, 17h.

The time in these notes is that of the Greenwich meridian.

NOVEMBER

The Planets

MERCURY is a morning object but will be too low for observation.

VENUS is -3.9 magnitude but rises little more than a hour before the Sun by the 30th. It should still be visible low in the SE in morning twilight.

MARS is 0.4 magnitude and in Capricornus setting before 22h throughout November. Moon to the south on the 21st. The position shown is for the 30th.

JUPITER is in Gemini and stationary on the 2nd, rising by 18h 30m by the 30th. Moon nearby on the 5th–6th.

SATURN is in Taurus, rising at sunset by the 30th. Moon close by on the 3rd and 30th.

URANUS is in Capricornus, setting before 22h by the 30th. Mars south of Uranus on the 26th. Moon nearby on the 22nd.

NEPTUNE is in Capricornus, setting about 20h 30m by the 30th. Mars to the south on the 4th. Moon nearby on the 20th.

The Moon

Full Moon 1d 06h
Last quarter 8d 12h
New Moon 15d 07h
First quarter 22d 23h
Full Moon 30d 21h

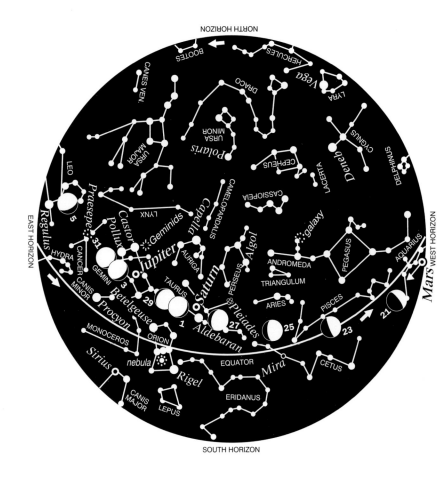

DECEMBER 1, 23h (11pm)

The aspect of the sky (apart from the Moon and Planets) will be approximately the same in other months at the following times:

September 1, 05h: October 1, 03h: November 1, 01h: January 1, 21h: February 1, 19h.

The time in these notes is that of the Greenwich meridian.

DECEMBER

The Planets

MERCURY is at superior conjunction on the 4th and then becomes an evening star but will not be observable.

VENUS is an evening star, setting an hour after the Sun on the 1st when it may be seen low in the SW but will be lost in twilight by mid-month.

MARS moves into Aquarius in early December and will be 0.7 magnitude by the end of the year. Moon to the south on the 20th.

JUPITER remains in Gemini, rising at sunset by the 31st a very bright -2.7 magnitude. Moon nearby on the 3rd and 30th.

SATURN is in Taurus and at opposition on the 3rd at -0.4 magnitude. It sets by 06h on the 31st. North of Aldebaran on the 17th. Moon nearby on the 1st and 28th.

URANUS is in Capricornus and sets by 20h by the 31st. Moon nearby on the 19th.

NEPTUNE is also in Capricornus setting by 18h 30m by the end of December. Moon nearby on the 18th.

The Moon

Last quarter 7d 20h
New Moon 14d 21h
First quarter 22d 21h
Full Moon 30d 11h

Eclipses on the 14th and 30th: *see* page 3
The Earth: Winter Solstice on 21d 19h

THE STARS

The stars are subdivided into magnitudes according to apparent brightness; the lower the number the brighter the star and the larger the dots on our monthly maps. Any star is about 2½ times as bright as one of the next magnitude. The faintest star ordinarily visible to the naked eye is of the 6th magnitude, or just one-hundredth of the brightness of one of the 1st, but that is possible only under a very clear sky. On a moonless night the total number of stars so visible is about 1,000. The faintest object detected with ground-based telescopes is of the 28th magnitude though the Hubble Space Telescope has now reached 30th magnitude.

Zero magnitude (0.0) represents a brightness 2½ times that of a standard first-magnitude star. Brightnesses in excess of this are indicated by a minus sign, the magnitude of Sirius, for example, being -1.47. Venus at its brightest is -4.6 or 145 times as bright as a first-magnitude star. The magnitude of the Full Moon is -12.5, equal to 250,000 first-magnitude stars. The stellar magnitude of the Sun is -26.6 or some 444,000 Full Moons.

The colours of the stars are indications of their surface temperatures. The temperature of a reddish star like Antares is about 3,000°C, and that of a bluish-white star, such as Vega, is about 11,000°C. The temperatures of orange, yellow and white stars are intermediate between these extremes.

An examination of the sky on a clear dark night shows that the distribution of stars is far from uniform. While there are distinct clusters of stars, such as the Pleiades and Praesepe, many other groupings consist of stars that just happen to lie in the same direction but at very different distances. The most noticeable concentration of stars is towards what we call the Milky Way, the faint band of light that passes through the following constellations: Puppis, Monoceros, Gemini, Auriga, Perseus, Cassiopeia, Cepheus, Cygnus, Aquila and Sagittarius. Not all of these constellations are above the horizon at any one time. The band of the Milky Way actually extends right round the sky passing through some southern constellations that never rise above the horizon in the British Isles.

Even binoculars show that the Milky Way is made up of thousands of stars, too faint to be seen with the naked eye. Our Sun is situated well away from the centre of a huge, flattened, disc-like system of stars 100,000 light years across called the Galaxy. It contains more than 100,000 million stars. When we look along the plane of the disc we see the star-clouds of the Milky Way; but when we look out above or below the plane we see far fewer stars.

From a distance, the Galaxy would look like that in Andromeda, visible to the naked eye only as a hazy oval patch of light. This is one of the nearer galaxies, only two-million light years away. Others have been found in their millions; some may be farther than 10,000 million light years distant, each containing thousands of millions of stars. The central bulge of our Galaxy lies towards the great star clouds in Sagittarius, not easily seen from our latitudes.

SUNSET, SUNRISE AND NAUTICAL TWILIGHT

Date		London Area				Edinburgh Area			
		Sunset	End NT	Begin NT	Sunrise	Sunset	End NT	Begin NT	Sunrise
Jan	1	16 00	17 20	06 45	08 08	15 45	17 22	07 08	08 44
	15	16 20	17 40	06 40	08 00	16 10	17 43	07 00	08 45
Feb	1	16 45	18 05	06 25	07 40	16 41	18 10	06 40	08 10
	15	17 15	18 30	06 00	07 15	17 16	18 38	06 15	07 38
Mar	1	17 35	18 50	05 35	06 50	17 47	19 05	05 40	07 03
	15	18 05	19 15	05 00	06 15	18 15	19 35	05 05	06 28
Apr	1	18 35	19 50	04 20	05 35	18 49	20 12	04 20	05 45
	15	19 00	20 10	03 40	05 00	19 20	20 52	03 45	05 05
May	1	19 25	20 50	03 05	04 30	19 50	21 35	02 45	04 25
	15	19 45	21 35	02 30	04 05	20 15	22 25	01 56	03 47
Jun	1	20 10	22 00	01 55	03 50	20 43	23 53	00 35	03 34
	15	20 20	22 29	01 33	03 40	21 00	----	----	03 25
Jul	1	20 25	22 25	01 40	03 45	21 01	----	----	03 31
	15	20 10	22 00	02 05	04 00	20 45	23 40	00 45	03 45
Aug	1	19 50	21 30	02 40	04 20	20 19	22 22	02 10	04 15
	15	19 25	20 50	03 15	04 45	19 50	21 30	03 00	04 42
Sep	1	18 50	20 10	03 50	05 10	19 09	20 42	03 45	05 15
	15	18 15	19 30	04 20	05 35	18 30	19 52	04 20	05 43
Oct	1	17 40	18 50	04 45	06 00	17 47	19 07	04 55	06 15
	15	17 05	18 20	05 10	06 25	17 05	18 32	05 25	06 40
Nov	1	16 35	17 50	05 40	06 50	16 31	17 57	05 52	07 19
	15	16 10	17 25	06 02	07 20	16 04	17 30	06 22	07 48
Dec	1	15 50	17 15	06 25	07 45	15 42	17 16	06 45	08 20
	15	15 50	17 13	06 37	08 03	15 36	17 10	07 00	08 38
	31	16 00	17 20	06 45	08 08	15 45	17 22	07 08	08 44

Times in UT. Also see notes overpage

29

Notes:

1. Times are given in Universal Time (= GMT): when British Summer Time (BST) is in force (usually from late March to late October) add 1 hour.

2. Nautical Twilight is defined as the moment when the Sun's true centre reaches a depression of 12 degrees below the horizon. Then it is dark enough to see the brighter stars and planets, and in suburban areas it often gets no darker. When no time is shown nautical twilight lasts all night.

3. Times given are approximate. They depend on the observer's latitude and longitude. Sunset, sunrise and twilight times will be 4 minutes earlier for every degree of longitude east of the Greenwich meridian and 4 minutes later for every degree west. London is on the Greenwich meridian; Edinburgh is about 3 degrees west or 12 minutes later.

4. The observer's latitude also affects these times: for example sunset occurs earlier in Edinburgh than London in winter but later in summer. It is not possible to cover more than two regions here, but a reasonable estimate can be made for other parts of the British Isles. The times may be used for any year. More explanation on this and other subjects can be found in *The Times Night Sky Companion*.

PRINCIPAL METEOR SHOWERS IN 2001

Name	Period of maximum visibility	Average hourly rate	Notes on visibility & moonlight
Quadrantids	2–4 Jan	10–100	Favourable. Moon 1st qtr 2nd. Radiant low in N in evening
Lyrids	21–22 Apr	10	Favourable. Moon new on 23rd
Perseids	11–14 Aug	60	Favourable before midnight. Last qtr on 12th.
Orionids	20–22 Oct	10–20	Favourable. 1st qtr on 24th
Taurids	late Oct–late Nov	5–10	Slow, from below Pleiades
Leonids	16–18 Nov	??	New on 15th Favourable, strong shower possible this year
Geminids	12–14 Dec	60	Moon new on 14th, favourable

Notes: The Leonids are normally a weak shower but every 33 years activity increases when the parent comet Tempel-Tuttle is near the Sun. Then high but usually short-lived activity may be seen as in the display of fireballs in 1998 and high rates of fainter meteors in 1999. The radiant, from where the meteors appear to come, is from within the 'sickle' (or reversed question mark) of the constellation Leo (major): *see* December chart (p.26).

Moonlight between first quarter and last quarter seriously interferes with the number of faint meteors seen. Fewer meteors are usually seen when the radiant area is low near the horizon. The radiant areas are shown on the monthly charts, except for the Taurids which covers a wide area and the Leonids which has not risen at the time of the November chart.

Further Events in 2001

Apart from eclipses and meteor showers, there are a number of events visible to the naked eye (some even better seen with binoculars) in 2001, the most striking of which are described here.

The monthly notes give the magnitudes of the planets as a guide to how bright they will appear (also *see* page 28). Venus and Jupiter are always so bright that they are easily identified but Mars and Saturn may be confused with nearby bright stars. Mercury, visible only in twilight, may sometimes be seen near a bright star. The following easily found stars, spread through the year, may help to identify the planets by comparing their brightness (magnitude): Polaris (2.0), Aldebaran (0.9), Sirius (-1.5), Castor (1.6), Procyon (0.4), Regulus (1.3), Arcturus (0.0), Spica (1.0), Antares (1.0) and Altair (0.8).

Mars and Antares. In March Mars will pass to the north of the bright star Antares in Scorpius, being at its closest on the 4th. Antares is a red giant star and at 1.0 magnitude the brightest star in the constellation of the Scorpion. Despite its considerable distance of 326 light years and relatively dull cool surface of around 3000°C, it appears bright to us because of its huge size. In early March Mars will be about 0.5 magnitude, rather brighter than Antares. About the beginning of nautical twilight (05h 30m) on 4 March, Antares will be in the south with Mars 5 degrees above it at an altitude above the horizon of about 15 degrees as seen from the southern British Isles. From farther north they will be nearer the horizon.

The name Antares is usually thought to derive from the Greek, meaning 'similar to Mars' or 'rival of Mars', referring to its reddish colour. This March will provide an opportunity to compare them.

Mars at opposition. Mars reaches a stationary point on 11 May in Sagittarius, then moves westwards (retrogrades) against the background of the stars. On 13 June Mars will be at opposition in Ophiuchus and due south at midnight and a very bright -2.4 magnitude. Mars will be stationary again on 19 July in Ophiuchus and then resumes eastward (direct) motion towards Sagittarius. Unfortunately for telescopic observers in the northern hemisphere, the closest approaches of Mars occur when the planet is south of the equator. This opposition at 68 million km Earth-Mars distance is favourable for southern observers. The closest possible distance at opposition is 56.3 million km and the farthest 100 million km. For reasons relating to the shape of Mars orbit minimum distance actually occurs on 21 June. At opposition Mars will reach a maximum of only 13 degrees above the southern horizon as seen from London and will appear even lower from farther north. More information on Mars and the other planets can be found in *The Times Night Sky Companion*.

Venus and Saturn. On 15 July Venus and Saturn will rise together in the northeast shortly before 02h and by the beginning of nautical twilight (02h 15m), when the brighter stars will begin to fade into the dawn sky, the planets will stand about 10 degrees above the eastern horizon. Venus will be a brilliant -4.1 magnitude and less than a degree above it at 0.2 magnitude will be Saturn. The waning crescent Moon will lie some 15 degrees to the right and Jupiter and Mercury will be just clearing the north-eastern horizon a similar distance to the left. Below Venus will be the bright star Aldebaran in Taurus.

Venus and Jupiter. On 6 August Venus will join Jupiter in Gemini for a fairly close approach, minimum distance just over a degree. About the beginning of nautical twilight (03h 00m in the south) Venus will be -4.0 magnitude and Jupiter just above it -2.0 magnitude. The two should make a striking pair. Some 10 degrees above and to the right will be Saturn near Aldebaran and to the left of Venus the bright stars Castor and Pollux.

Venus and Regulus. On 20 September about the beginning of nautical twilight (04h 15m), Venus (-4.0 magnitude) will pass close to the north of the bright star Regulus (1.3 magnitude) in Leo Major. They will be about 10 degrees above the eastern horizon. Above and to the right will be Jupiter in Gemini and above Jupiter, Saturn in Taurus.

Occultation of Saturn. As the Moon advances eastwards through the constellations each month, it passes in front of stars and less often planets. These events are called lunar occultations. Most stars occulted are faint and in some years there may be no planetary occultations at all. On 3 November 2001 the waning gibbous Moon will pass in front of the planet Saturn. The event takes place at a convenient time, about 21h–22h (9–10pm), when the Moon will be 30 degrees above the eastern horizon. Each hour the Moon advances by its own diameter (half a degree) and the duration of an occultation is greatest when the star or planet passes centrally behind the Moon; then it can last as long as an hour. But the duration may be only minutes or even seconds if the star or planet just grazes the Moon's northern or southern limb.

As this event will take place after full Moon, Saturn will disappear at the bright limb (left edge) of the Moon at about 21h and will reappear about an hour later at the dark (right) limb. The exact timing depends on the observer's position on the Earth's surface so it is as well to begin looking half an hour before 21h, when Saturn will be easier to see in the Moon's glare, and then watch the two bodies as they draw closer together. The Moon is so bright that it is often difficult to see a star or planet very close to the illuminated limb of the Moon, so in this case the reappearance at the dark limb may be easier to see: much depends on the clarity of the sky.

In the case of a star the disappearance and reappearance is instantaneous and without detailed information it is difficult to focus on the exact point of reappearance. A planet can take many seconds to disappear or reappear, so it should be quite easy to pick it up again. The use of binoculars, preferably supported on a tripod, is strongly recommended. More precise timings will be given in *The Times* monthly Night Sky notes nearer the time.